Exploration of Venus

Patrick H. Stakem

(c) 2021

Table of Contents

Introduction..3
Author..5
The Planet Venus...7
Life in the Clouds..8
    Exploring Venus..8
        Pioneer Venus 10
        Magellan..11
        Mariner-1..12
        Parker Solar Probe..12
        Vega Missions...12
        Galileo...13
        Cassini...13
        Messenger...14
        BepiColombo..14
        Venus Express..14
        Planned Missions..15
Bibliography..19
Resources...22
Glossary of terms...24
If you enjoyed this book, you might also be interested in some of these..27

Introduction

Venus is a reeking hell-hole of a planet. It is about the same size as the Earth, and is the next planet closer to the Sun. It's atmosphere has significant concentrations of sulfuric acid, with a bit of hydrofluoric acid. The temperatures and density's of the clouds are incredible. Venus has no magnetic field, nor does it have a moon.

Venus was known to pre-history, including the Babylonians, Greeks, Chinese and the Mayans. It was well studied by Galileo, who observed the phases of Venus. The Romans named it after their goddess of love and beauty.

A cylinder seal from BC 3100-2900 BC records observations of the planet. The Sumerians knew it was both the morning and evening star. It was representative of the goddess Ishtar. From a Babylonian library there are recording of the appearances of Venus spanning 21 years. Venus is sometimes visible during the day.

Venus was the most important celestial body to the Maya. It was called Chac Ek, the great star. The Maya thought that Venus and the stars influenced life on Earth. They developed a highly accurate 584 day religious calendar based on Venus and other celestial body's.

At certain times, Venus can shine bright enough to cast a shadow on Earth. It has no moons. Venus is roughly the size of Earth, and is a rocky planet, like Earth.

Venus rotates retrograde or clockwise, opposite of the Earth and all the other planets except Uranus. Before observations were made by radio telescopes in 1906, Venus was thought to have a lush Earth-like environment. It is the hottest planet, in spite of the fact that Mercury is closer to the Sun. That is because the dense atmosphere traps heat. Venus is the victim of a runaway greenhouse effect.

There is some controversy about whether Venus has lightning in its atmosphere. Venus Express did discover an ozone layer. There is extensive volcanism. Nearly 1,000 craters are distributed across the Venetian surface. One of the larger ones is 174 miles in diameter. The dense atmosphere slows incoming meteors quite a bit. Some are slowed to a crawl in the dense atmosphere, and do not produce an impact crater.

Venus has a magnetic field, but it is much smaller than Earth's. This means that Venus interacts more with the solar wind.

A prime meridian was defined for Venus, to allow for placement of surface features. It currently passes through the central peak of the crater Aridne.

Except for the extreme temperatures, Venus could be Earth's twin. It was the second visited by a spacecraft, after Earth. That was by Mariner 2 in 1962. Mariner was heading to the sun, but got 42 minutes of data about

Venus as it went past. Actually, about 30 miles above the surface, the temperatures and pressures are very Earth-like. If you weren't in clouds of sulfuric acid, it would seem like a nice day.

In 2020, observations by the James Clerk Maxwell telescope and the Atacama Large Millimeter Array (Chile) showed the presence of large amounts of Phosphine (PH3), in the carbon dioxide atmosphere. There was more than could be accounted for by non-biological sources. This chemical was located some 48 kilometers above the surface. This is not necessarily proof of biological organisms on Venus, but presents a puzzle. It is a highly toxic respiratory poison to us, and is produced by microbial life on Earth. An alternative is the sensor had detected sulfur dioxide from the numerous volcanoes. The latest approach to figuring out what is going on uses the Atacama Large Array radio telescope in Chile.

Venus has been explored by Earth based radar, starting with the Goldstone radio telescope facility in 1961.

Author

Mr. Patrick H. Stakem has been fascinated by the space program since the Vanguard launches were televised in 1957. He received a Bachelors degree in Electrical Engineering from Carnegie-Mellon University, and Masters Degrees in Physics and Computer Science from the Johns Hopkins University. At Carnegie, he worked with a group of undergraduate students to re-assemble and operate a surplus missile guidance computer, which was later donated to the Smithsonian. He was brought up in the mainframe era, and was taught to never trust a computer you could lift.

He began his career in Aerospace with Fairchild Industries on the ATS-6 (Applications Technology Satellite-6), program, a communication satellite that developed much of the technology for the TDRSS (Tracking and Data Relay Satellite System). He followed the ATS-6 Program through its operational phase, and worked on other projects at NASA's Goddard Space Flight Center including the Hubble Space Telescope, the International Ultraviolet Explorer (IUE), the Solar Maximum Mission (SMM), some of the Landsat missions, and Shuttle. He was posted to NASA's Jet Propulsion Laboratory for MARS-Jupiter-Saturn (MJS-77), which later became the *Voyager* mission, and is still operating and returning data from outside the solar system at this writing. He initiated and lead the Flight Linux Project for NASA's Earth Sciences Technology Office. During a career as a NASA support contractor from 1971 to 2013, he worked at almost all of the NASA Centers.

He has the NASA Shuttle Program Manager's Commendation Award as well as the NASA Apollo-Soyuz Test Program Award, and two NASA Group Achievement Awards. He has worked on the SS Freedom Project.

Mr. Stakem was affiliated with the Whiting School of Engineering of the Johns Hopkins University he supported the Summer Engineering Bootcamp Projects at Goddard Space Fight Center for 2 years. He developed and presented Cubesat courses. He also taught for Lolola University in Maryland and Capital University.

The Planet Venus

The Venus environment has proven extremely hostile. It seems our sister world, next towards the Sun from us, is in a environmental runaway condition. Heavy greenhouse clouds trap the solar energy, and cause massive global warming on a planetary scale. The surface temperature is high enough to melt some metals. This is very hard on computers, and electronics in general. We need to find out what sent wrong on Venus, and try to avoid that on Earth.

Venus' atmosphere is 96% carbon dioxide at a surface pressure of nearly 100 times Earth's, a greenhouse gone wild. It has no moons. Venus is roughly Earth-sized, but something went terribly wrong. It also has clouds of sulfuric acid, that landers have to get through. There is not mush of a magnetic field, but there is active volcanism. Venus has more volcano's than any other planet in our solar system, some 1,600 known. More than half the planet is covered by a lava plain. There are more than 1,000 volcanic structures known. There are 1,600 volcanoes known, but these might all be extinct. The Magellan mission saw evidence of recent activity at the highest volcano, Maat Mons. Most volcanos are of the shield type. Volcanic activity has not been seen, but there are large amounts of sulfur dioxide in the atmosphere. In the middle cloud layer, winds can reach 450 miles per hour.

Life in the Clouds

The surface of Venus is too hot to allow for life, but life may have evolved in the dense cloud layer. Water, carbon dioxide and sunlight are plentiful in the clouds. These are essential for photosynthesis. The clouds are also known to contain water vapor at their bottoms. Did life evolve there, and what was it like? We don't know.

Exploring Venus

Since the 1960's, there have been forty-six spacecraft launches to Venus. Not all of them were successful.

The Soviet space program sent a series of probes to Venus named Venera. Early efforts were either crushed in the dense atmosphere, or suffered thermal damage. Tyazhely Sputnik failed to leave Earth orbit. Venera-1's telemetry failed 7 days after launch as it was enroute. Venera's 3-6 were very similar. Venera-2 was the first man-made object to impact an other's planet surface. It crash landed without sending any data. This was not going to be easy. Venera-4 was the first spacecraft to explore the atmosphere of another planet. It was about 100 Earth atmospheres of pressure at the surface. Venera-5 and -6 were repurposed to take readings of the atmosphere as long as they could. They descended by parachute until their battery's failed. The Venera-7 was redesigned, and the mission had a goal of surface sample return. It struck the surface harder than planned, but returned temperature data for about 20 minutes. The Venera-8 probe returned data for some 50 minutes. It showed temperatures of 869 degrees F at a surface pressure of 90 Earth Atmospheres. Venera-9 was redesigned again. It operated for 53 minutes at the surface, and managed to send back photos. Venera-10 lasted 65 minutes at the surface. Venera-11 lasted 95 minutes, but the lens caps failed to release. Venera-12 lasted 110 minutes, but, again, the lens caps stuck.

The Soviet Union landed 10 probes on Venus' surface. The longest survivor lasted two hours. This is perhaps because the surface temperature is now known to be greater tha the melting point of solder. The shortest duration was 23 minutes.

Venera-13 and -14 returned color photos of the surface. Venera lasted for 127 minutes, and V-14 lasted 57 minutes. Not giving up, Venera-15 and -16 were repurposed as orbiters, with surface imaging imaging radar. Further Soviet and US efforts involved observation from Venus orbit. Russia's mission to Halley's comet, VeGa, used the Venera design. Venera-D was never built.

There was a series of Russian spacecraft sent to explore Venus. Number 1, Tyazhely Sputnik, experienced a premature upper stage cutoff, and never left Earth orbit. Venera-1 had a communications failure. Unit 2MV-1 suffered a premature third stage cut-off in 1962. Unir 2MV-2 had an upper stage fuel valve problem that caused the engine to fail to ignite. 3MV-1 number 2 had a leak that caused the propellant in the fuel lines to freeze, in

1964.

Kosmos-27 experienced a upper stage attitude control failure, and never got out opf Earth orbit. Zond-1 got to fly by Venus in 1964, but the electronics failed, and there was no communication with Earth. Kosmos-96 had a third stage combustion chamber explosion in 1965. Kosmos-167's upper stage failed to ignite in 1967. Kosmos-359 had a launch failure, and did not leave Earth orbit in 1970. The same was the case for Kosmos-482.

U.S. Mission Mariner-5 did a successful flyby in 1967. Mariner-10 launched in 1973, and did a flyby in 1974. It then continued on to Mercury.

Venus Express, an ESA mission, is in Venus Polar orbit. It found a massive double atmospheric vortex (storm) at the south pole. Venus Express operated from 2005-2014. Venus has no moons, but does have Trojans. The Japanese Venus Climate Orbiter "Akatsuki" was launched in 2010, but failed to achieve Venus orbit. It orbited the Sun for 5 years, and finally got into Venus orbit in 2015.

Venus differs from every other planet in our Solar System in that it rotates in the opposite direction of the other planets. The length of a Venusian year is about 225 Earth days, and its rotation period is 243 days.

Pioneer Venus

This 1978 U. S. mission had two spacecraft, the Orbiter and the Multiprobe. Unit number 1 entered orbit successfully in 1978. This was a project of NASA-Ames. The multiprobe launched four small probes into the atmosphere. They sent back data on the atmosphere, and one survived the surface landing for an hour. There were seventeen science instruments, including a radar mapper. The spacecraft was built by Hughes Aircraft. Unit two, the multi-probe, entered the Venus atmosphere in December of 1978.

The orbiter weighted in at more than a thousand pounds, and had 17 different instruments.

It was in a 66,900 km orbit. It entered the Venusian atmosphere and was destroyed in 1992.

Pioneer Venus was able to observe Halley's Comet from Venus orbit, when the comet was not visible from Earth.

Magellan

This one ton spacecraft launched in May of 1988 from the Space Shuttle, and entered Venetian orbit in 1990. It carried a synthetic aperture radar. It continued to operate until it was de-orbited in 1994. Interestingly, to save costs, the spacecraft was built mostly from flight space parts from other spacecraft. The flight computer was from Galileo, as was the command and data system. Thrusters and main body came from the Voyager program. Voyager also supplied the high gain antenna. There were two solar arrays, providing 1,200 watts of power to charge the batterys. Magellan could store 225 megabytes of data for later transmission to Earth on digital tape recorders.

The radar operated in three modes, synthetic aperature, altimetry, and radiometry. The spacecraft was doing well, so the mission was extended from May 1991, to January 1992. A map of 95% of the surface was constructed from Magellan's data.

Mariner-1

Mariner-1 was launched in 1962 on a trip to Venus, but it failed to reach orbit. Mariner-2 did a fly-by in December of the same year. Unit number two had an upper stage fuel valve fail to open, and it stayed in Earth orbit.

Parker Solar Probe

The Parker solar probe, on its way to the Sun, made multiple fly-bys of Venus, with a closest encounter of 515 miles in 2020. The probe had flown through Venus' atmosphere, as evidenced by the spacecraft's detection of a natural radio signal.

Vega Missions

The Vega Missions were developed by the Soviet Union, with the participation of eight other country's. These missions also took advantage of the near flyby of Halleys comet as well.

Vega 1 and 2 were identical spacecraft, weighting in at 4,920 kg. They carried 11 instruments. Vega 1 and 2 arrived at Venus in June of 1985. Each had a detachable lander, and a balloon.

Vega two survived landing and sent relevant data. The atmospheric pressure at the landing site was 91 Earth atmospheres, with the temperature being 865 degrees F.

Electronics in general does not work well in this environment.

The balloons, termed aerobots were targeted to float 34 miles above the surface. They survived and returned data for 46 hours.

The two spacecraft, no longer operational, continue to orbit the Sun.

Galileo

The Galileo Mission launched on the shuttle Atlantis in 1989, heading to Jupiter. In 1990 it observed Venus on a fly-by. Wait, isn't that going the wrong way? Well, it was taking advantage of a maneuver called gravity assist or slingshot, which makes the trip longer, but uses less fuel. The first use of a gravity assist maneuver was in 1959 by the Soviets. It's now common to use a Venus assist to get missions to Mars and the outer planets.

Even though its primary mission was Jupiter, but it did collect data on Venus during the flyby. It arrived at Jupiter in 1995.

During the Venus flyby, it used some of its instruments that had not been to Venus before. One of these was NIMS, near infrared mapping spectrometer. Others were the Ultraviolet spectrometer, and the Energetic particles detector. Lightning on Venus was confirmed by the Plasma wave detector.

Cassini

The Cassini-Huygens mission launched in 1997, aboard

a Titan IV headed for Saturn. The mission was joint among NASA, ESA, and the Italian Space Agency. Huygens was the lander. The spacecraft also used two Venus fly-by's for an assist to the realm of the outer planets. It captured data at Venus as it swung around the planet. It went on to spend 13 years at Saturn, returning much useful data.

Messenger

The Messenger spacecraft was headed to Mercury, the next planet before the Sun. It was launched in 2004, and went by Venus twice in 2006 and 2007, taking useful data. On the first fly-by, the Sun was in the way, preventing data to be sent back. On the Second fly-by the ESA Venus Express was orbiting Venus and simultaneous measurements were possible

BepiColombo

This Venus explorer is a joint effort of ESA and Japan Aerospace Exploration Agency (JAXA). It's primary mission is as a Mercury orbiter, but it did two Venus flybys in 2020 and 2021. It was trying to detect any phosphine in the atmosphere, but there was a concern the MERTIS (Mercury Radiometer and Thermal Infrared Spectrometer) instrument was not sensitive enough.

Venus Express

Venus Express was the first ESA spacecraft to explore Venus. It launched in 2005 on a Soyuz, and achieved a nine-day polar orbit around Venus. Its design generally followed that of the the Mars Express. It's mission was to gather data on the Venetian atmosphere dynamics. The mission recorded the Venusian surface temperatures. The mission continued to provide good data until November of 2014, when evidently all the propellant was exhausted.

The instruments included ASPERA-4, the Analyzer of Space Plasma and Energetic Atoms. This was looking at the interaction of the Venusian atmosphere and the solar wind. A magnetometer was used to measure Venus' magnetic field. It turns out to be very weak. It also carried a wide field camera for imaging volcanic activity and airglow. A Planetary Fourier Spectrometer observes in the infrared, and does vertical sounding of the atmosphere. An imaging spectrometer characterizes the atmosphere. The ionosphere is studied by passing radio waves through the atmosphere for reception and analysis back on Earth.

An interesting imaging experiment was used to observe signs of life on Earth. We know the answer, but we need to know how well we can tell that at a distance. The spacecraft's radio signal was lost in January of 2015.

Planned Missions

Even with as many missions that have studied Venus, more are under development. The private company Rocket Lab is working on a Venus atmospheric probe for 2023.The company is an American aerospace company, with a wholly owned subsidiary in New Zealand.

ISRO's Shukrayaan-1 is an orbiter and atmospheric balloon, to be launched in 2024. It's focus will be surface and sub-surface stratigraphy,

NASA has an orbiter in development for a 2028 launch. The acronym stands for Venus Emissivity, Radio Science, Interferometric synthetic aperture radar (InSAR), Topography, and Spectroscopy.

NASA is also working on the DAVINCI atmospheric probe for a 2029 launch. The acronym means Deep Atmosphere Venus Investigation of Noble gases, chemistry, and imaging plus. It will have both an orbiter and a descent probe. It will provide the first images of the surface since 1981. It will have a mass spectrometer, a tunable laser spectrometer, and a multispectral camera for both the UV and the Near IR bands.

Roscosmos is working on Venera-D, an orbiter and lander, for launch in 2029.

Envision is an ESA orbiter project, to be launched in the early 2030's. It will use a high resolution synthetic aperture radar for mapping of the surface, and the atmosphere. There is also a radar sounder, and 3-channel spectroscopy instrument.

Rocket Labs, of California, is designing a new satelite bus, named Photon. This will be launched to Venus in 2023. It will support a laser-tunable mass spectrometer in the Venus atmosphere,

Proposed missions, not yet fully defined, include:

CUVE is a 12-U cubesat mission to explore Venus in the Ultraviolet. This is being done for NASA by the University of Maryland. It will have a multi-spectral UV imager, a high resolution ultraviolet spectrometer, and small UV telescope.

HAVOC is the High Altitude Operational Concept addresses a crewed mission to Venus. Observations would be conducted from orbit, or lighter-than-air craft. At Venus, the mission duration would be 30 days. It would take 300 to get home. A follow-on would put astronauts into the atmosphere for 1 Earth year. Further along a Venus orbiting Space Station is envisioned.

VAMP is an atmosphere balloon mission for 2029. VICI and Visage are NASA lander missions for 2027. VOX is a proposed orbiter for 2027. Zephyr is a Venus rover mission considered for 2039.

In 2021, NASA announced two new Venus missions for the time frame of 2028-2030. These are DAVINCI and Veritas. Davinci stands for Deep Atmosphere Venusd Investigation of Noble Gases. Veritas is also an acronym, Venus emissivity Radio Science InSar, Topography and Spectroscopy. ESA announced their Envision mission, with a NASA provided Synthetic Aperature Radar, VenSAR.

Wrap-up

Venus got the attention of many of the ancient peoples, as it did not act like the other stars in the sky. It was definitely a wanderer. The Maya and the Sumerians kept careful records of its rising and settings, and could predict its position in the future. Did they think it was the abode of other beings? Was it like Earth? The Mayan name for the planet is Chac ek. They used careful ovservations of the planet to develop a highly accurate calander.

On of the early Russian landers failed abruptly while characterizing the surface temperature. Turns out, it is higher than the melting point of solder, and all of the electronics failed.

But we know that Venus is out closest neighbor, and could be Earth like if we could terraform it to get rid of those cloud layers. Could it be harboring life now, in those cloud layers? There's a lot more to know and explore.

Despite the many missions to collect data at Venus, we don't really know uch about our neighbor.

Can it be terra-formed? Should it?

Bibliography

Aguilar, David A. Space Encyclopedia: A Tour of Our Solar System and Beyond (National Geographic Kids), 2013, ISBN-1426309481.

Bezard, Brun et al (ed) *Venus III: The View After Venus Express,* 2020 ed, ISBN-978-9402419351.

Bova, Ben *Venus,* Grand Tour Book 6, 2001, ASIN-B003GWX8P8.

Burgess, Eric *Venus, An Errant Twin*, 1985, ISBN-978-0231058568.

Cattermole, Dr. Peter Venus: The Geological Story, 1994, ISBN-978-0801847875.

Filberto, Justin, et al, "Present-day Volcanism on Venus as evidenced from weathering rates of olivine, 2020, https://www.ncbi.nlm.nih.gov/pmc/articles/PMC6941908/

Ford, Kevin S. *Optimizing aerobot exploration of Venus*, 1994, ASIN-B00HQ7IJQ2.

Ford, J. et al Guide to Magellan Image Interpretation, JPL Publication 93-24, 1993.

Goldstein, Margaret J. Discover Venus, 2018, ASIN-1541523401.

Grinspoon, David Harru, *Venus Revealed: A New Look Below The Clouds Of Our Mysterious Twin Planet*, 1998, ISBN-978-0201328394.

Hunten, D. (et al) *The Planet Venus,* (The Planetary Exploration Series), 1998, ISBN-978-0300049756.

Keller, Raymond A. *Venus Rising: A Complete History of the Second Planet*, 978-1882658312.

Malcuit, Robert J. *The Twin Sister Planets Venus and Earth: Why are they so different?,* 1992, ISBN-978-3319113876.

Marov, Mikhail *The Planet Venus* (The Planetary Exploration Series) First Edition, ISBN-978-0300049756.

Miller, Ron *Venus,* ISBN 978-0-7613-2359-4. 2015, ISBN-

Ottewell, Guy, *Venus a Longer view*, 2020, ISBN-0934546812.

Roth Ladislav E., Wall, Stephen D. The Face of Venus: The Magellan Radar Mapping Mission, 2012, ISBN-:978-1478350637.

Taylor, Frederic W. *The Scientific Exploration of Venus*, 2014, ISBN-978-1107023482

Wulf, Andrea *Chasing Venus: The Race to Measure the Heavens, 2012,* ASIN B0067TGUQQ.

Ya, Michail et al, *The Planet Venus*, 1998, ISBN-978-0300049756.

Zharkov, V. N. (et al) *Venus Geology, Geochemistry, and Geophysics: Research Results from the Soviet Union*, 1992, ISBN-978-0816512225.

Resources

- https://www.nasa.gov/mission_pages/
- www.planetary.org
- Goldstein, Margaret J. *Discover Venus*, 2018, ISBN-978-1541523401.
- Venus Fact sheet, NASA, https://nssdc.gsfc.nasa.gov/planetary/factsheet/venusfact.html
- Venus, Profile, https://web.archive.org/web/20150906034051/http://solarsystem.nasa.gov/planets/venus
- Venus crater database, https://www.lpi.usra.edu/resources/vc/vchome.shtml
- Venus Profile, https://web.archive.org/web/20150906034051/http://solarsystem.nasa.gov/planets/venus
- http://solarsystem.nasa.gov/missions/profile.cfm?Sort=Target&Target=Venus&MCode=Pioneer_Venus_01&Display=ReadMore
- https://nssdc.gsfc.nasa.gov/planetary/pioneer_venus.html
- https://en.wikipedia.org/wiki/Venus_tablet_of_Ammisaduqa
- google.com/maps/space/venus
- https://nssdc.gsfc.nasa.gov/planetary/planets/venuspage.html
- google.com/maps/space/venus

- *Magellan at Venus,* J. of Geophysical Research, Vol. 97, no. E8 and E10, A.G.U., Washington, D.C., 1992.
- Young, C. (ed) *Magellan Venus Explorer's Guide*, JPL Publication 90-24, 1990, Online.
- Fimmel, R. et al *Pioneer Venus*, NASA SP-461, Washington, D.C., 1983.
- Pioneer Venus Special Issue, J. Geophysical Research, Vol. 85, December, 1980.
- wikipedia, various.

Glossary of terms

Apo- – the point furthest away from a primary by a body that orbits it.
ASIN – Amazon Standard Inventory Number.
Asteroid – a chunk of rock; a minor planet.
AU – astronomical unit, mean distance from the Earth to the Sun, 93,000,000 miles.
CME – Coronal Mass Ejection – burst of plasma and magnetic fields from a sun's corona.
Coma – Comet's tail
Comet – a solar system object consisting of ice, dust, and gas, in highly eccentric orbit.
Ecliptic – the apparent path that the Sun seems to follow, the same as the Earth's orbit.
Equinox – 2 days per year when there are equal periods of daylight and darkness.
ESA – European Space Agency.
EU – European Union.
IAU – International Astronomical Union.
IR – infrared.
ISRO – Indian Space Development Organization.
Heliocentric – sun-centered.
HST – Hubble Space Telescope.
IAU – International Astronomical Union.
ISBN – International Standard Book Number.
JPL – NASA's Jet Propulsion Lab, Pasadena, California
Light year – the distance light travels in one year. 9.5 $\times 10^{12}$ kilometers.
Mbps – 10^6 bits per second.

Mbytes – mega (10^6) bytes.
Moon – smaller astronomical body in orbit of a planet.
NASA – National Aeronautics and Space Administration.
NEO – near Earth object.
Nephelometer – instrument to measure concentration of suspended   particles in a liquid or gas.
NGST – Next Generation Space Telescope – renamed after James Webb.
Nova – transient astronomical event involving a bright new star   that fades over time.
NSF – (U.S.) National Science Foundation.
OCPP – Cloud photo-polarimeter.
OEFD – electron temperature.
OIMS – ion mass spectrometer.
OIR Onfrared radiometer.
OMAG – magnetometer.
ONMS – neutral mass spectrometer.
OPA – (solar wind) plasma analyzer.
ORAD – Surface Radar Mapper.
ORPA – charged particle retarding potential analyzer.
Orbit – the path of one body around another, that are linked by   gravity.
OUVS – airglow ul;traviolet spectromerer
Planet – a body orbiting a star.
Red Shift – an apparent shift of electromagnetic radiation toward an increasing wavelength due to the doppler effect.
SMM – Solar Maximum Mission.
Solar flare – a sudden rapid emission of electrons, ions, and atoms from a star.
Solar System – A star and its associated planets and such.

Solar wind – stream of charged particles emitted from a star's upper atmosphere.
Solstice – day of the shortest or longest period of daylight.
Stratigraphy - study of rock stratas.
StScI – Space Telescope Science Institute (JHU)
Tidal lock – where the same side of a object always faces the primary it is orbiting.
UV - ultraviolet
Venera – series of space probes for the Planet Venus.

If you enjoyed this book, you might also be interested in some of these.

Stakem, Patrick H. *16-bit Microprocessors, History and Architecture*, 2013 PRRB Publishing, ISBN-1520210922.

Stakem, Patrick H. *4- and 8-bit Microprocessors, Architecture and History*, 2013, PRRB Publishing, ISBN-152021572X,

Stakem, Patrick H. *Apollo's Computers,* 2014, PRRB Publishing, ISBN-1520215800.

Stakem, Patrick H. *The Architecture and Applications of the ARM Microprocessors,* 2013, PRRB Publishing, ISBN-1520215843.

Stakem, Patrick H. *Earth Rovers: for Exploration and Environmental Monitoring,* 2014, PRRB Publishing, ISBN-152021586X.

Stakem, Patrick H. *Embedded Computer Systems, Volume 1, Introduction and Architecture*, 2013, PRRB Publishing, ISBN-1520215959.

Stakem, Patrick H. *The History of Spacecraft Computers from the V-2 to the Space Station*, 2013, PRRB Publishing, ISBN-1520216181.

Stakem, Patrick H. *Floating Point Computation*, 2013, PRRB Publishing, ISBN-152021619X.

Stakem, Patrick H. *Architecture of Massively Parallel Microprocessor Systems*, 2011, PRRB Publishing, ISBN-1520250061.

Stakem, Patrick H. *Multicore Computer Architecture*, 2014, PRRB Publishing, ISBN-1520241372.

Stakem, Patrick H. *Personal Robots*, 2014, PRRB Publishing, ISBN-1520216254.

Stakem, Patrick H. *RISC Microprocessors, History and Overview*, 2013, PRRB Publishing, ISBN-1520216289.

Stakem, Patrick H. *Robots and Telerobots in Space Applications*, 2011, PRRB Publishing, ISBN-1520210361.

Stakem, Patrick H. *The Saturn Rocket and the Pegasus Missions, 1965*, 2013, PRRB Publishing, ISBN-1520209916.

Stakem, Patrick H. *Visiting the NASA Centers, and Locations of Historic Rockets & Spacecraft*, 2017, PRRB Publishing, ISBN-1549651205.

Stakem, Patrick H. *Microprocessors in Space*, 2011, PRRB Publishing, ISBN-1520216343.

Stakem, Patrick H. Computer *Virtualization and the Cloud*, 2013, PRRB Publishing, ISBN-152021636X.

Stakem, Patrick H. *What's the Worst That Could Happen? Bad Assumptions, Ignorance, Failures and Screw-ups in Engineering Projects, 2014,* PRRB Publishing, ISBN-1520207166.

Stakem, Patrick H. *Computer Architecture & Programming of the Intel x86 Family, 2013,* PRRB Publishing, ISBN-1520263724.

Stakem, Patrick H. *The Hardware and Software Architecture of the Transputer*, 2011, PRRB Publishing, ISBN-152020681X.

Stakem, Patrick H. *Mainframes, Computing on Big Iron*, 2015, PRRB Publishing, ISBN- 1520216459.

Stakem, Patrick H. *Spacecraft Control Centers*, 2015, PRRB Publishing, ISBN-1520200617.

Stakem, Patrick H. *Embedded in Space,* 2015, PRRB Publishing, ISBN-1520215916.

Stakem, Patrick H. *A Practitioner's Guide to RISC Microprocessor Architecture*, Wiley-Interscience, 1996, ISBN-0471130184.

Stakem, Patrick H. *Cubesat Engineering*, PRRB Publishing, 2017, ISBN-1520754019.

Stakem, Patrick H. *Cubesat Operations*, PRRB Publishing, 2017, ISBN-152076717X.

Stakem, Patrick H. *Interplanetary Cubesats*, PRRB Publishing, 2017, ISBN-1520766173 .

Stakem, Patrick H. Cubesat Constellations, Clusters, and Swarms, Stakem, PRRB Publishing, 2017, ISBN-1520767544.

Stakem, Patrick H. *Graphics Processing Units, an overview*, 2017, PRRB Publishing, ISBN-1520879695.

Stakem, Patrick H. *Intel Embedded and the Arduino-101, 2017,* PRRB Publishing, ISBN-1520879296.

Stakem, Patrick H. *Orbital Debris, the problem and the mitigation*, 2018, PRRB Publishing, ISBN-*1980466483*.

Stakem, Patrick H. *Manufacturing in Space*, 2018, PRRB Publishing, ISBN-1977076041.

Stakem, Patrick H. *NASA's Ships and Planes*, 2018, PRRB Publishing, ISBN-1977076823.

Stakem, Patrick H. *Space Tourism*, 2018, PRRB Publishing, ISBN-1977073506.

Stakem, Patrick H. *STEM – Data Storage and Communications*, 2018, PRRB Publishing, ISBN-

1977073115.

Stakem, Patrick H. *In-Space Robotic Repair and Servicing*, 2018, PRRB Publishing, ISBN-1980478236.

Stakem, Patrick H. *Introducing Weather in the pre-K to 12 Curricula, A Resource Guide for Educators*, 2017, PRRB Publishing, ISBN-1980638241.

Stakem, Patrick H. *Introducing Astronomy in the pre-K to 12 Curricula, A Resource Guide for Educators*, 2017, PRRB Publishing, ISBN-198104065X.
Also available in a Brazilian Portuguese edition, ISBN-1983106127.

Stakem, Patrick H. *Deep Space Gateways, the Moon and Beyond*, 2017, PRRB Publishing, ISBN-1973465701.

Stakem, Patrick H. *Exploration of the Gas Giants, Space Missions to Jupiter, Saturn, Uranus, and Neptune*, PRRB Publishing, 2018, ISBN-9781717814500.

Stakem, Patrick H. *Crewed Spacecraft*, 2017, PRRB Publishing, ISBN-1549992406.

Stakem, Patrick H. *Rocketplanes to Space*, 2017, PRRB Publishing, ISBN-1549992589.

Stakem, Patrick H. *Crewed Space Stations,* 2017, PRRB Publishing, ISBN-1549992228.

Stakem, Patrick H. *Enviro-bots for STEM: Using Robotics in the pre-K to 12 Curricula, A Resource Guide for Educators,* 2017, PRRB Publishing, ISBN-1549656619.

Stakem, Patrick H. *STEM-Sat, Using Cubesats in the pre-K to 12 Curricula, A Resource Guide for Educators*, 2017, ISBN-1549656376.

Stakem, Patrick H. *Lunar Orbital Platform-Gateway*, 2018, PRRB Publishing, ISBN-1980498628.

Stakem, Patrick H. *Embedded GPU's*, 2018, PRRB Publishing, ISBN- 1980476497.

Stakem, Patrick H. *Mobile Cloud Robotics*, 2018, PRRB Publishing, ISBN- 1980488088.

Stakem, Patrick H. *Extreme Environment Embedded Systems,* 2017, PRRB Publishing, ISBN-1520215967.

Stakem, Patrick H. *What's the Worst, Volume-2*, 2018, ISBN-1981005579.

Stakem, Patrick H., *Spaceports*, 2018, ISBN-1981022287.

Stakem, Patrick H., *Space Launch Vehicles*, 2018, ISBN-1983071773.

Stakem, Patrick H. *Mars*, 2018, ISBN-1983116902.

Stakem, Patrick H. *X-86, 40th Anniversary ed*, 2018, ISBN-1983189405.

Stakem, Patrick H. *Lunar Orbital Platform-Gateway*, 2018, PRRB Publishing, ISBN-1980498628.

Stakem, Patrick H. *Space Weather*, 2018, ISBN-1723904023.

Stakem, Patrick H. *STEM-Engineering Process*, 2017, ISBN-1983196517.

Stakem, Patrick H. *Space Telescopes,* 2018, PRRB Publishing, ISBN-1728728568.

Stakem, Patrick H. *Exoplanets*, 2018, PRRB Publishing, ISBN-9781731385055.

Stakem, Patrick H. *Planetary Defense*, 2018, PRRB Publishing, ISBN-9781731001207.

Patrick H. Stakem *Exploration of the Asteroid Belt*, 2018, PRRB Publishing, ISBN-1731049846.

Patrick H. Stakem *Terraforming*, 2018, PRRB Publishing, ISBN-1790308100.

Patrick H. Stakem, *Martian Railroad,* 2019, PRRB Publishing, ISBN-1794488243.

Patrick H. Stakem, *Exoplanets,* 2019, PRRB Publishing, ISBN-1731385056.

Patrick H. Stakem, *Exploiting the Moon,* 2019, PRRB Publishing, ISBN-1091057850.

Patrick H. Stakem, *RISC-V, an Open Source Solution for Space Flight Computers,* 2019, PRRB Publishing, ISBN-1796434388.

Patrick H. Stakem, *Arm in Space*, 2019, PRRB Publishing, ISBN-9781099789137.

Patrick H. Stakem, *Extraterrestrial Life*, 2019, PRRB Publishing, ISBN-978-1072072188.

Patrick H. Stakem, *Space Command*, 2019, PRRB Publishing, ISBN-978-1693005398.

CubeRovers, A Synergy of Technologys, 2020, PRRB Publishing, ISBN-979-8651773138.

Robotic Exploration of the Icy moons of the Gas Giants. 2020, PRRB Publishing, ISBN- 979-8621431006

Hacking Cubesats, 2020, PRRB Publishing, ISBN-979-8623458964.

History & Future of Cubesats, PRRB Publishing, ISBN-979-8649179386.

Hacking Cubesats, Cybersecurity in Space, 2020, PRRB Publishing, ISBN-979-8623458964.

Powerships, Powerbarges, Floating Wind Farms: electricity when and where you need it, 2021, PRRB Publishing, ISBN-979-8716199477.

Hospital Ships, Trains, and Aircraft, 2020, PRRB Publishing, ISBN-979-8642944349.

<u>2020/2021 Releases</u>

CubeRovers, a Synergy of Technologys, 2020, ISBN-979-8651773138

Exploration of Lunar & Martian Lava Tubes by Cube-X, ISBN-979-8621435325.

Robotic Exploration of the Icy moons of the Gas Giants, ISBN- 979-8621431006.

History & Future of Cubesats, ISBN-978-1986536356.

Robotic Exploration of the Icy Moons of the Ice Giants, by Swarms of Cubesats, ISBN-979-8621431006.

Swarm Robotics, ISBN-979-8534505948.

Introduction to Electric Power Systems, ISBN-979-8519208727.

Centros de Control: Operaciones en Satélites del Estándar CubeSat (Spanish Edition), 2021, ISBN-979-8510113068.

www.ingramcontent.com/pod-product-compliance
Lightning Source LLC
Chambersburg PA
CBHW030043230526
45472CB00005B/1654